SEE INSIDE

AN OIL RIG AND TANKER

Series Editor: **R. J. Unstead**

Warwick Press
New York/London/Toronto/Sydney
1988

Series Editor
R. J. Unstead

Author
Jonathan Rutland

Illustrations
Doug Harker Doug Post
Ian Robertson Ken Rush
Jim Stanes Michael Treganza
Mike Roffe Paul Wright

Published in 1988 by Warwick Press,
387 Park Avenue South, New York, New York 10016.
First published in 1978 by Hutchinson & Co.
(Publishers) Ltd.
Revised edition published 1988 by Kingfisher Books.

Printed in Hong Kong.

5 4 3 2 1

ISBN 0-531-19046-3
Library of Congress Catalog Card No. 88-50109

CONTENTS

The Story of Oil

Crude oil was formed millions of years ago from countless tiny plants and creatures living in the seas, lakes, and rivers. Dead bodies of these plants and creatures sank to the sea bed and were buried by thick layers of mud which had been washed off the land. The lower layers hardened into rocks because of the pressure of the immense weight above them. Within these rocks crude oil and natural gas formed from the remains of the animals and plants. They are called fossil fuels.

Some rock is *porous*—the grains of which it is composed have small interconnected spaces between them rather like a sponge. Other rock layers are *impermeable*—nothing can pass through them. Under the pressure of newer rocks forming above them the oil and gas was squeezed out of the rocks where it was formed and moved to the porous rocks above. It moved to the porous rocks above. It moved sideways and upward until its path was blocked by impermeable rock. The result was an *oil trap*.

Geologists searching for oil, study the way the rocks lie. This gives them an idea of where oil may be trapped underground. When they find areas which look likely, they carry out what is called a *seismic survey*. They fire explosive charges in shallow holes in the ground; these charges shake the rock and the geologists measure the time it takes for the resulting shock waves to be bounced back from the layers of different sorts of rock. From this they get a good idea of the shape of the rocks thousands of feet beneath the surface. From the *seismic survey* the geologists fix a likely spot to drill an exploration well, called a *wildcat*. In the past most oil wells were on land. But fossil fuels are so valuable that today oil companies are drilling for oil and gas from reservoirs in rocks deep under the seabed.

Above: A concrete gravity production platform being constructed. This type of platform is often used in deep water instead of the piled steel platform described on page 6. The finished platform is towed out to the oil field (opposite page). Then it is sunk into position on the seabed. It will rest there safely because it is so heavy.
Below: A layer of porous oil-bearing rock (blue) between two layers of impermeable rock. The oil (red) has been trapped in a reservoir formed when the earth moved. Millions of years ago the oil-bearing rock layer was on the surface of the earth, and the oil formed from plants and bodies of small creatures that lived then.

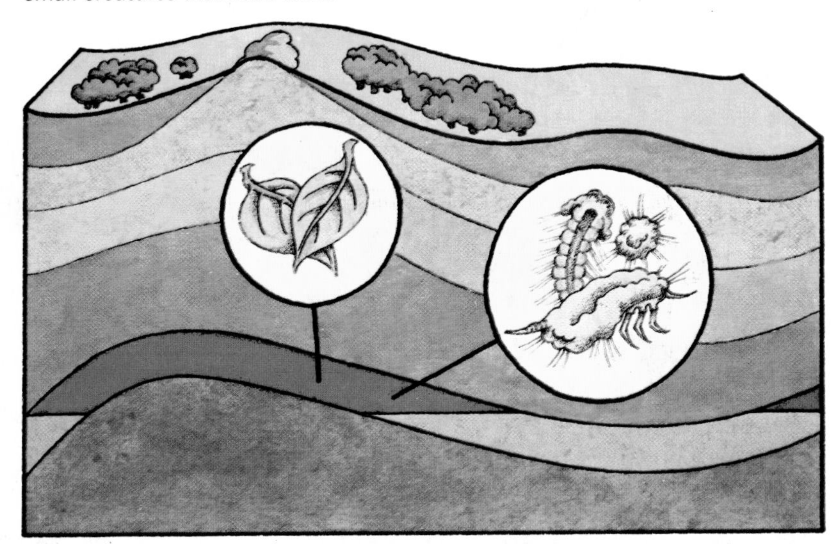

A self-elevating or jack-up rig, used instead of a floating rig in water up to 300 feet deep. It is towed into position with its legs raised (left). They are then lowered until they rest firmly on the seabed and the platform is lifted above the waves (above).

The Drilling Rig

The semi-submersible platform (right) is an exploration rig. It floats, steadied by bouyancy tanks, and is positioned using anchors and computer-controlled motors. Its job is to drill a test well or borehole, but for deeper water a drill ship is used.

On land or sea, the biggest part of the drilling gear is a tall metal tower, the *derrick*. Hanging down inside this is the *kelly*. This is a length of square pipe which fits through a square hole in the *rotary table* at the foot of the derrick.

When work begins, the kelly is first hoisted out of the way, and a round drill pipe is lowered through the derrick floor. On the bottom of this drill pipe is the drill bit which cuts through the rock. A second pipe is screwed to the first, the two are lowered, and a third added—and so on until the drill almost reaches the seabed. Then the top drill pipe is screwed to the kelly. The whole *drill string* (kelly, drill pipes, and bit) is lowered until the foot of the kelly is locked in the rotary table. At last drilling can begin. An engine turns the rotary table which turns the kelly, drill pipes, and bit. The drill cuts into the rock. Soon another length of pipe must be added. The engine is stopped, the kelly unscrewed and the new pipe fitted.

During drilling a special kind of mud is pumped down the drill pipes. This *oils* the bit and washes the chips of rock cut by the drill out of the borehole. Scientists examine the chips, and can judge from them if the borehole is getting near oil. If there is oil, it may gush up and be wasted. So a valve called a *blow-out preventer* is fixed by remote control at the top of the borehole.

Above: The drill bit has wheels with sharp teeth. When the drill pipe turns, the wheels bore into the rock.

Production Platform

When the drilling rig has found a good supply of oil it raises anchor and moves on to drill another test well somewhere else. Usually more than one exploration well is drilled in an area before a *production platform* is built and floated out. You can see a finished production platform on the left. It stands on the seabed on a huge steel frame called a *jacket*. The feet of the jacket are anchored to the seabed by piles. These are rather like giant nails nearly 6 feet thick. They are hammered deep down (150 to 300 feet) into the rock and hold the platform steady, even in stormy seas. The deck of the production platform is then built on top of the jacket. above the waves. There are two basic types: steel or concrete. The picture shows a "piled steel" platform.

Whichever kind of platform is used, its first job is to drill production wells. These are made in the same way as they are by the exploratory drilling rig (page 4), but instead of one borehole there are up to 40 or 50. The drill is made to slant out as it cuts down into the rock, so that wells fan out beneath the platform. Thus one platform can gather oil from a wide area.

Once the wells have been drilled, the oil can be piped up to the platform. The flow of oil is regulated by a group of pipes, taps, and valves called a *Christmas tree*—so named because it looks a little like one, with all its branching pipes and taps. Special equipment is needed to remove any water and gas in the oil. Sometimes the gas is *flared-off* (burned in a flare). But if there is enough gas it is piped ashore to add to our supplies of natural gas. It is also used to power generators on board the platform.

Production platform on the sea bed. The whole structure may be over 650 feet tall, and 100 or more people may live aboard it.

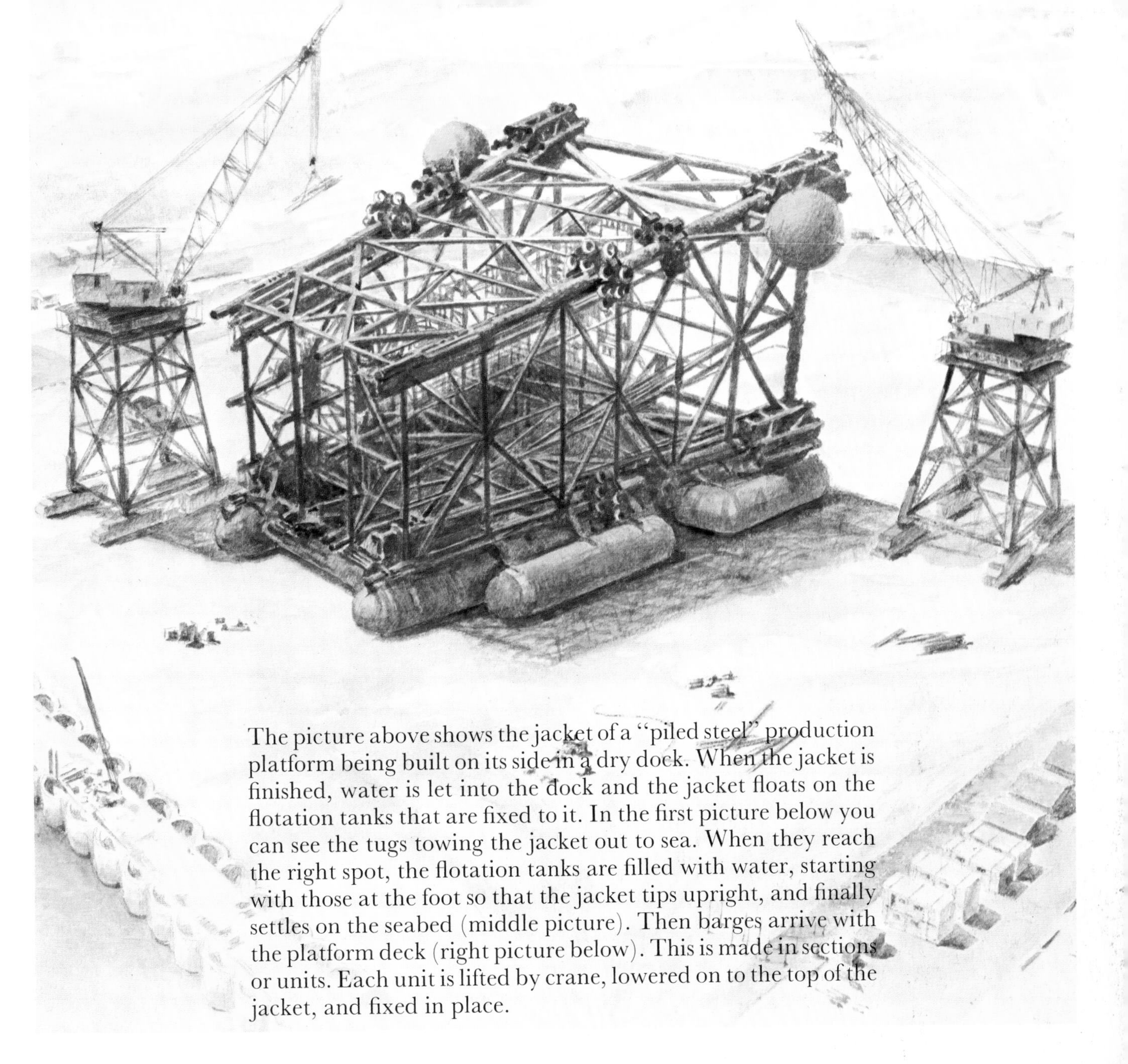

The picture above shows the jacket of a "piled steel" production platform being built on its side in a dry dock. When the jacket is finished, water is let into the dock and the jacket floats on the flotation tanks that are fixed to it. In the first picture below you can see the tugs towing the jacket out to sea. When they reach the right spot, the flotation tanks are filled with water, starting with those at the foot so that the jacket tips upright, and finally settles on the seabed (middle picture). Then barges arrive with the platform deck (right picture below). This is made in sections or units. Each unit is lifted by crane, lowered on to the top of the jacket, and fixed in place.

Living on an Oil Platform

Above: One of the few exciting moments in an oilman's day, as the helicopter arrives with the mail, papers, and new crew members. Landing the helicopter can be tricky and dangerous, and in fog or high wind impossible. Most supplies arrive in special oil rig supply ships. Unloading in calm weather is easy, but in rough seas the rig's crane operator needs split-second timing to lift supplies safely onto the rig.

Below: Meal time in the canteen. The food is usually very good indeed. Outside working hours and meal times there is little to do except play cards or darts, watch the weekly film, or sleep. The cabins are neat and clean, but small. Few people choose to work on an oil rig for fun. They do so because it is better paid than most jobs ashore.

Noise is the first thing you would notice if you visited an oil platform—especially one in the North Sea where high winds and rough seas can be deafening. On top of this there is the roar of the flare, the heavy throb of electricity generators, and the whine of the drilling gear. Then there is the smell—of oil, of the special drilling *mud*, and of fumes from the diesel generators. There is the coldness and wetness, and the blast of the wind. The decks are often wet, oily, and slippery, and the wind can blow you off your feet. Guardrails are fitted to stop people being blown overboard.

The work is tough too. Most of the crew work 12 hours a day for 14 days on end, and then have two weeks off ashore. Crew members range from the *roustabouts*—odd job men whose work includes scraping off the ever present rust, painting, and washing the decks—to highly trained divers and geologists.

The actual drilling team is led by a man called the *tool pusher*. Working for him are the *drillers* and *rotary helpers*. They fit new pipes to the drill string. High up above, the *derrickman* works lifting gear to pick up the next pipe. On the derrick floor one man swings the pipe over, and two others help him screw it into the pipe below with a huge hydraulic wrench (picture right). Their toughest work comes when the drill bit needs replacing. The whole drill string, which may be $2\frac{1}{2}$ miles or more in length, must be raised, unscrewed and stacked, the new bit fitted, and the whole lot put back together again. The drill string must also be drawn up when the borehole is "cased." Heavy steel tubes (the casing) are lowered and cemented in the borehole to strengthen it and stop it caving in. The *mud men* look after the mud pumped down the drill string.

On a floating exploration rig there is a *barge master*, who is in charge of the rig's seaworthiness; and the *ballast man*. He works in a cabin in one of the legs (page 5), seeing that the amount of water (ballast) in the pontoons is just right to keep the platform at the right level.

Bringing the Oil Ashore

An oil tanker or VLCC (Very Large Crude Carrier) moored to and collecting oil from a massive underwater oil storage tank. This one is shaped rather like an upsidedown funnel, but others are drum shaped. The tanks are built on shore, floated out and sunk in position by filling them with sea water. When production begins, oil flowing into the tank forces the water out.

The crude oil is usually pumped ashore along a pipeline on the seabed. This is a long string of pipes welded together and coated with tar and cement, linking the production platform to an oil storage terminal or refinery (see page 20) on the shore. From there the oil—or the products made from it, such as gasoline and kerosene—may be sent along pipelines across the land, or it may travel in huge ships called tankers.

Laying a pipeline under water (illustrated opposite) is a difficult and very expensive job. It can cost as much to lay a 30-inch wide pipeline as it does to build a freeway. Some oil fields do not produce enough oil to make a pipeline worthwhile. Others are too far from land. In these cases tankers go out to the platform to collect the oil and bring it back to a refinery.

Tankers are much too large and heavy to tie up safely alongside a platform. Instead a mooring buoy is anchored in the sea nearby. It is called a SBM, or Single Buoy Mooring. The tanker ties up to just one buoy. When wind or currents move the ship, it can swing safely around the buoy. Oil from the platform is piped to the buoy, and the tanker collects oil from it, or from the storage tanks below the buoy as in

the picture. You can see the flexible pipe, floating on the sea, which carries the oil from the buoy to the tanker.

If the oil comes straight up from the well to the platform and on to the buoy, the flow of oil must be stopped when there is no tanker. This wastes valuable production time. So some platforms and some SBMs have huge built-in oil storage tanks. On a few fields there are gigantic storage tanks which rest on the seabed. You can see one kind in the picture below.

The buoy or storage tank to which the tanker is moored is usually about a mile away from the production platform, and is connected to it by an underwater pipeline. This distance gives the huge unwieldy tanker plenty of room to swing around the buoy and to approach it from any direction as the wind and currents change.

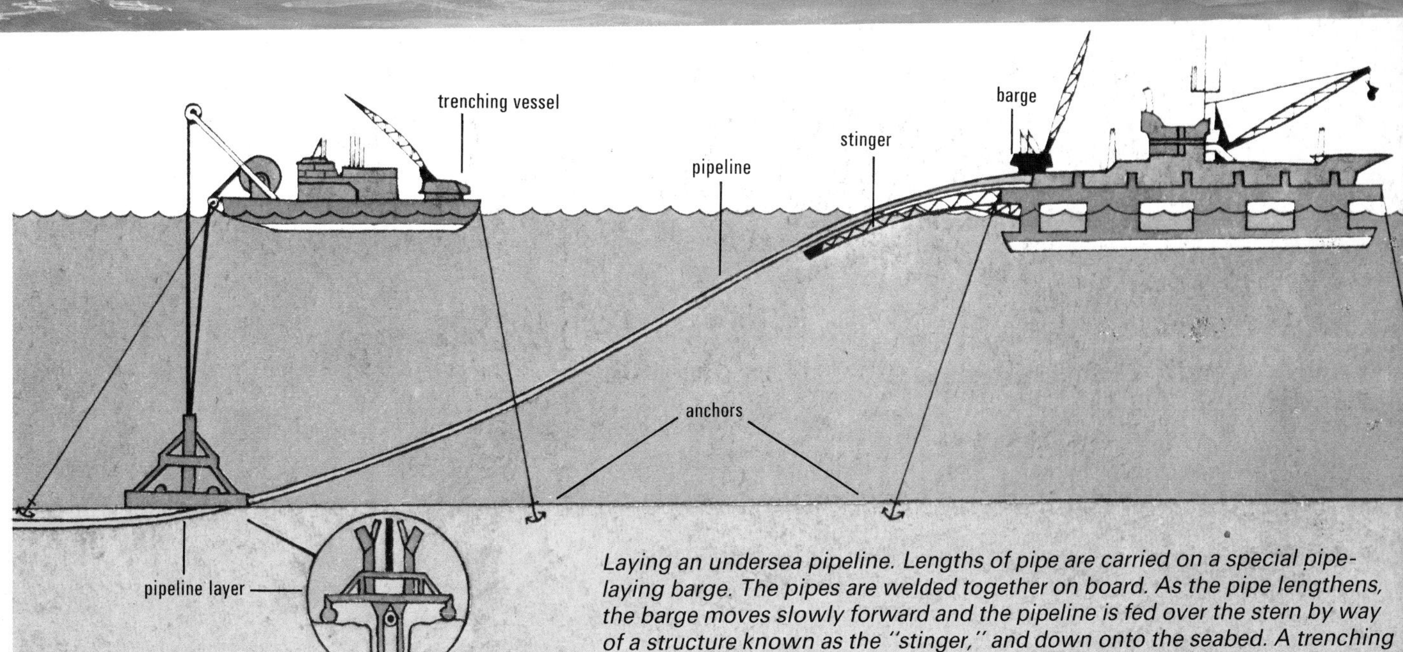

Laying an undersea pipeline. Lengths of pipe are carried on a special pipe-laying barge. The pipes are welded together on board. As the pipe lengthens, the barge moves slowly forward and the pipeline is fed over the stern by way of a structure known as the "stinger," and down onto the seabed. A trenching vessel follows and digs a trench in which the pipeline rests.

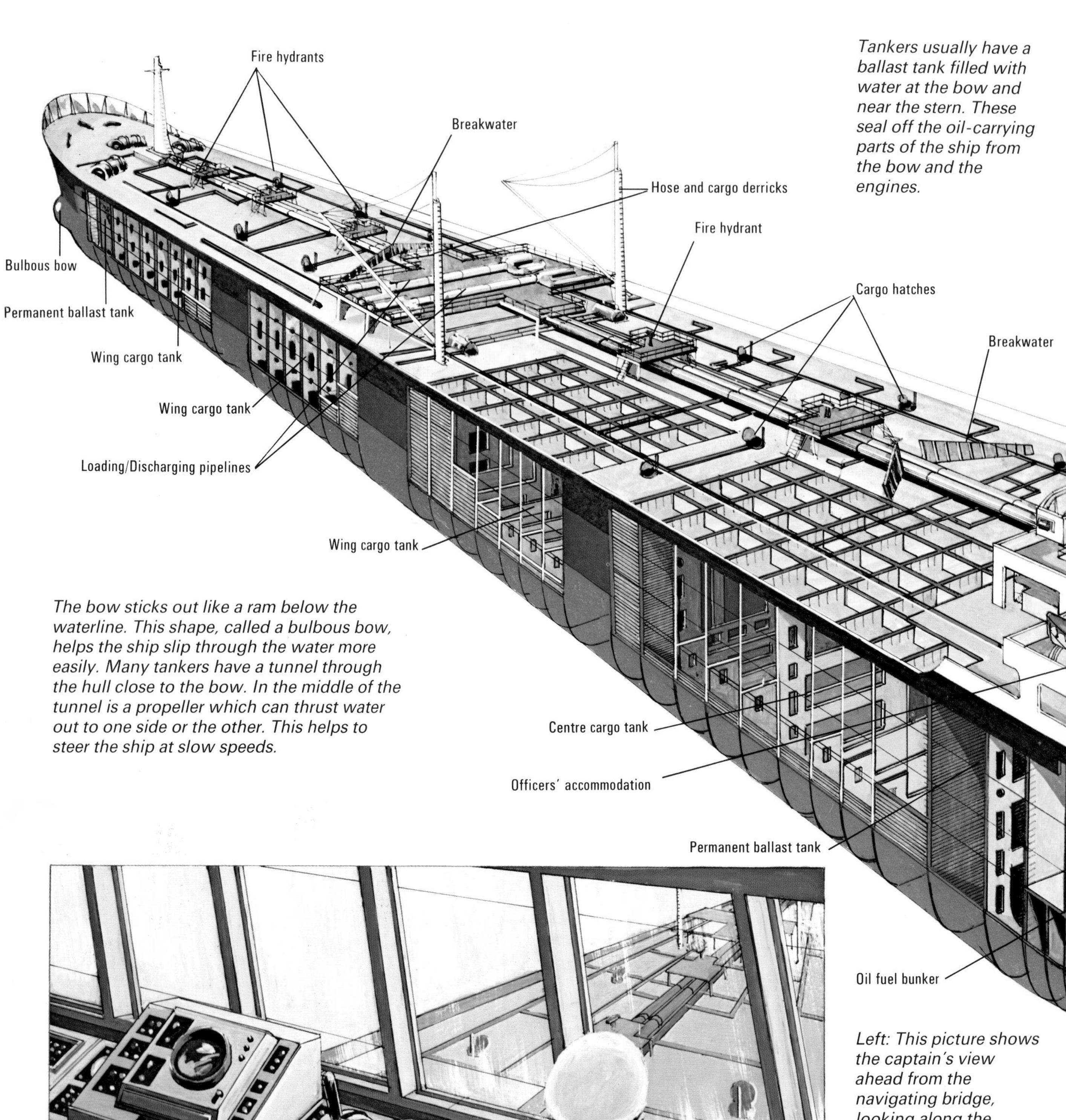

Tankers usually have a ballast tank filled with water at the bow and near the stern. These seal off the oil-carrying parts of the ship from the bow and the engines.

The bow sticks out like a ram below the waterline. This shape, called a bulbous bow, helps the ship slip through the water more easily. Many tankers have a tunnel through the hull close to the bow. In the middle of the tunnel is a propeller which can thrust water out to one side or the other. This helps to steer the ship at slow speeds.

Left: This picture shows the captain's view ahead from the navigating bridge, looking along the enormous length of the tanker. It gives an idea of how difficult these great ships are to steer and control. You can see some of the electronic instruments and controls. These include a "collision avoidance system"—a radar screen which shows the captain the course of his ship and of any others nearby.

Inside an Oil Tanker

Most tankers carry crude oil only, but some are designed to carry liquified natural gas (LNG), which must kept cold. They collect their load from the terminal or production area, take it to the shore refinery, and return empty for more. The engines, crew's quarters, control rooms, and navigating bridge are all at the stern (back) of the ship. There they are safely away from the dangerous cargo—the oil itself catches fire easily enough, but the fumes from it are far more dangerous and explosive.

The oil-carrying space is split into a number of separate tanks. This means that various kinds of oil can be carried without mixing (crude oil differs slightly from field to field). It also means that the oil does not slop about too much in rough weather, and that if the ship is holed in an accident the entire cargo will not pour out.

In calm seas the crew can ride bicycles up and down the deck. But in rough weather waves break over the deck. Then a crew member with a job to do near the bow walks along a raised gangway.

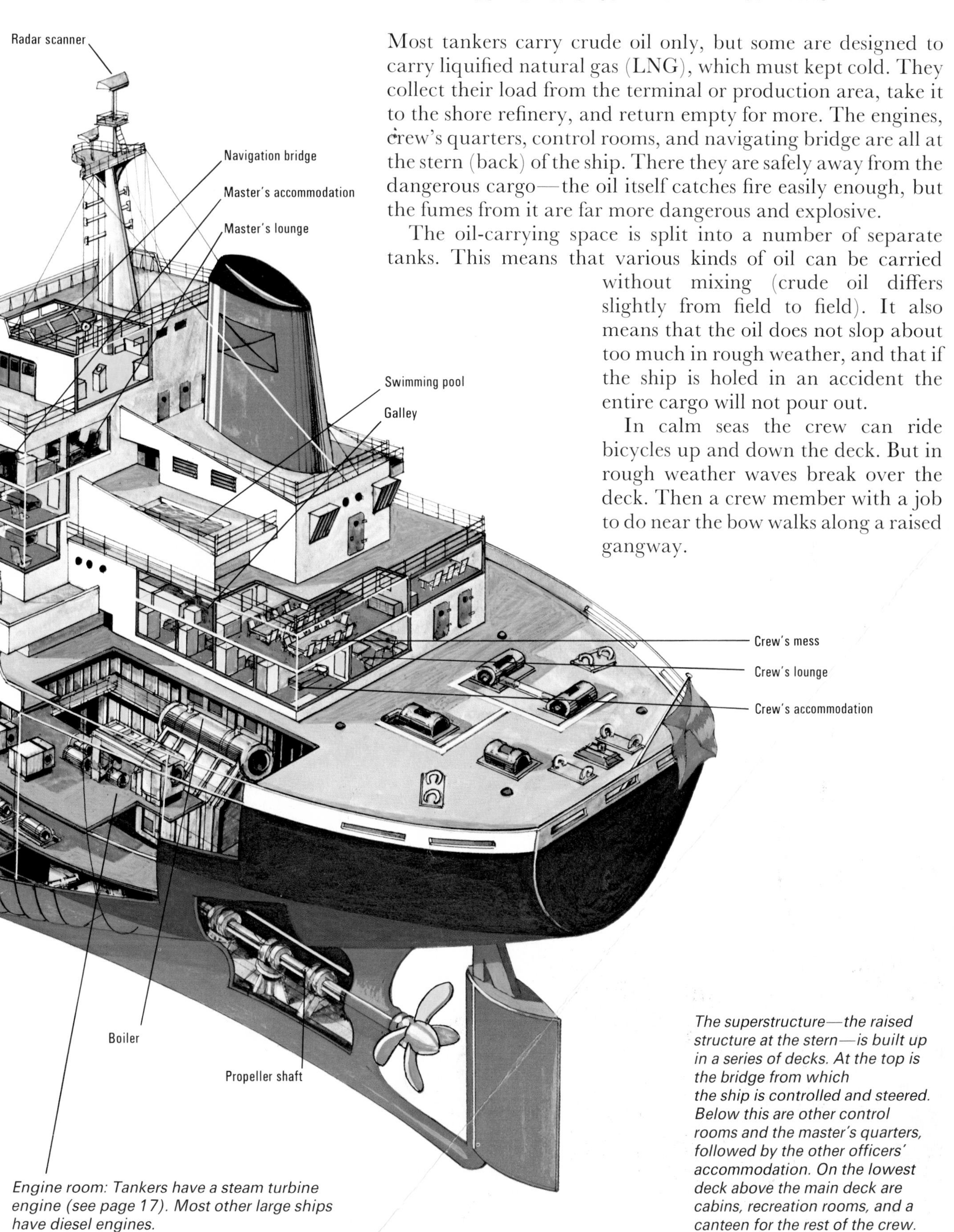

Engine room: Tankers have a steam turbine engine (see page 17). Most other large ships have diesel engines.

The superstructure—the raised structure at the stern—is built up in a series of decks. At the top is the bridge from which the ship is controlled and steered. Below this are other control rooms and the master's quarters, followed by the other officers' accommodation. On the lowest deck above the main deck are cabins, recreation rooms, and a canteen for the rest of the crew.

Building a Tanker

Tankers, like oil rigs, are usually built in a dry dock. This is a huge concrete "basin" at the water's edge, cut off from the water by massive watertight gates. When the ship is finished, the gates are opened to let the water in, and the ship floats out. Smaller ships are made on a sloping slipway. They are launched by sliding them down the slipway into the water. This is not possible with very long vessels such as tankers. Half way down the slipway the bow would rest on land and the stern in the water. The middle would be hanging in the air, and the ship would bend or break its "back." But as dry docks are very expensive, some tanker builders make their ships in two separate sections. Each section is built and launched on a slipway, and the two halves are joined together when they are afloat.

The traditional way of building ships was to lay down the keel (the "backbone") and to build the hull from the keel upward, finally fixing on the covering of steel plates. All the work was done outside at the slipway or dry dock. Today, large sections of the hull are made under cover in a large assembly shop. When each section is ready, cranes lift it out to the dry dock and hold it in place while workmen weld it onto the ever-growing hull. In some shipyards even this job is done under cover. Tracks link the assembly shop and the dock. The stern section is built first, on the track inside the workshop. It is then pushed along the track until only the edge to be

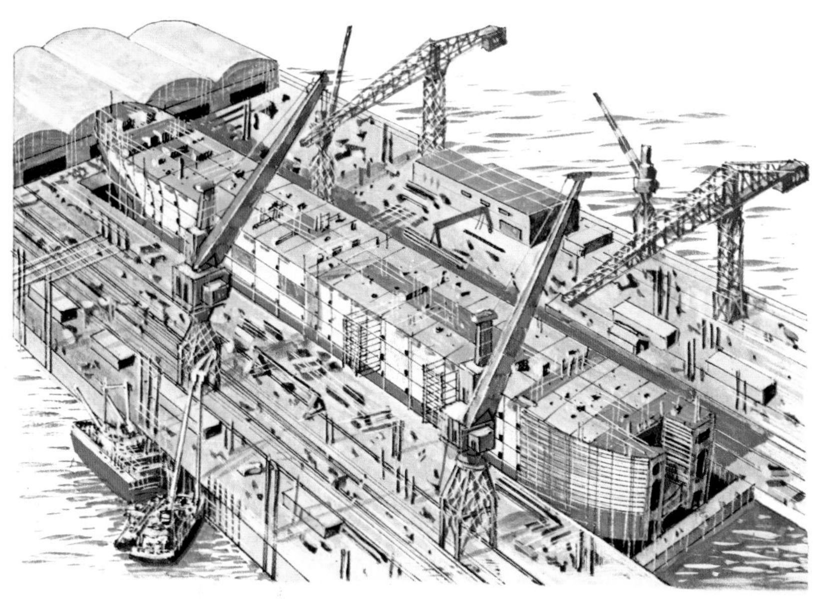

Top and center: The stern part of the tanker (top) nears completion. When it is ready the middle section (center) and then the bow will be welded into place.

Left: The lower part of the bulbous bow is lowered into position. Sections like this are built under cover in the assembly shed.

Above: Welders at work joining another section to the tanker's ever-growing hull.

Right: A giant supertanker being built in a dry dock. Soon the mass of cranes and scaffolding will be removed and the ship will be floated out for her trials.

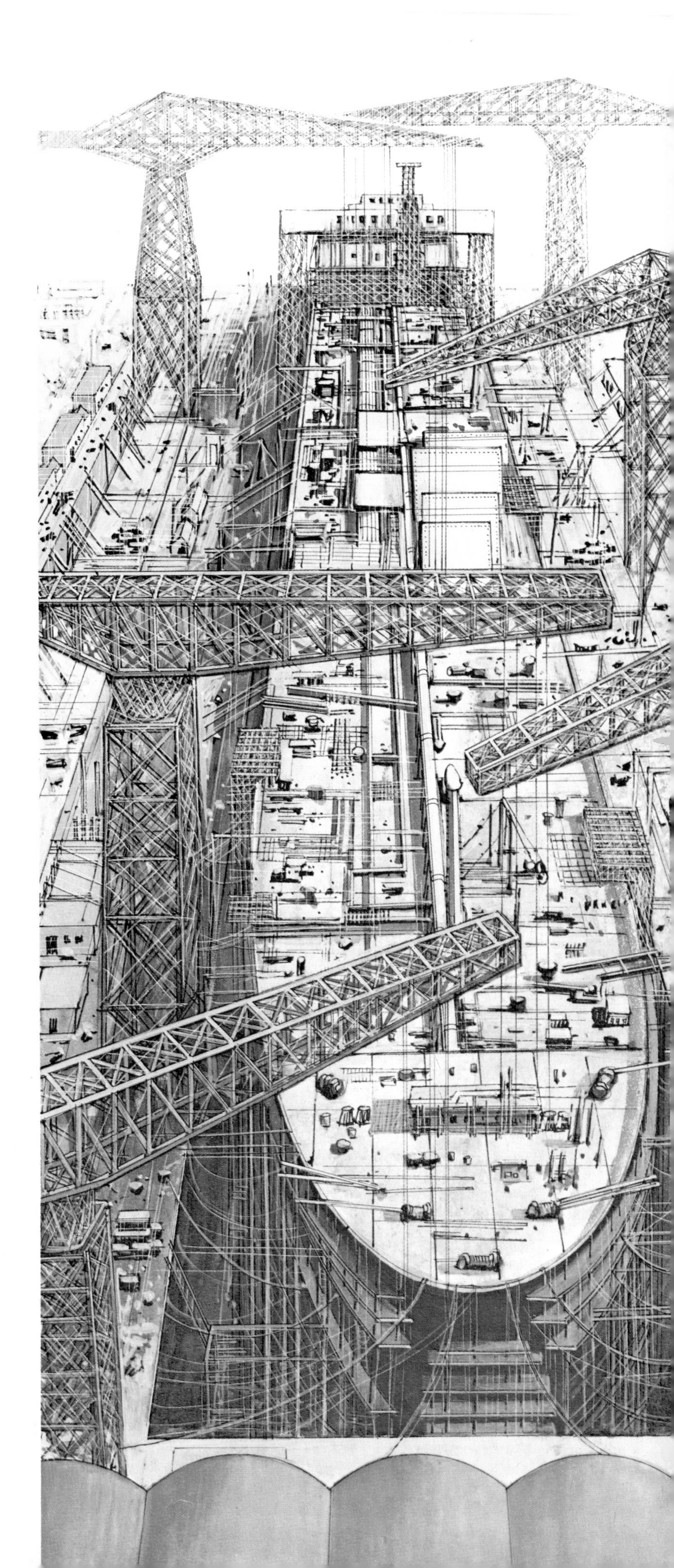

joined to the next section is left inside. The next section is made, joined to the stern unit, and pushed out—and so on.

Fitting out

When the hull is complete, the tanker is ready for fitting out. Engine, boilers, and other machinery are lowered and fixed in place, followed by the superstructure on top of the main deck, the rudder and propeller, and countless other details. At last the tanker is ready for her trials, when tests are made on her speed and seaworthiness, and on how easy she is to steer and control. No two ships are quite the same, and it is vital to know, for example, how quickly the rudder turns the ship at different speeds, how the ship is affected by wind and waves; and so on.

THE WORLD'S BIGGEST SHIPS

Tankers are the biggest ships ever, larger even than aircraft carriers. They are weighed in deadweight tonnage (dwt), the tonnage needed to bring down the ship from its unloaded height above the water to its load-water line. So dwt is the ship's carrying capacity. The world's largest tanker is the *Seawise Giant*, with a dwt of 564,763, but smaller tankers are now preferred.

The Tanker at Sea

Even the largest tankers have a crew of only around 40 men. This is one reason why tankers are so big. They do not need a larger crew than smaller ships, so one huge tanker can carry oil more cheaply than several small ships. The master and his crew have all sorts of electronic equipment to help them navigate and run the ship. Once at sea an automatic pilot holds the ship on course. It is linked to the compass and the rudder. Radio navigation equipment receives signals from special transmitters around the world and works out from them the ship's exact position. In case this goes wrong, tankers like all other large ships carry instruments for working out position from the sun—they are a sextant, for measuring the height of the sun above the horizon, and a chronometer (a very accurate clock). Then there are radar sets which help the navigating officer to see ahead at night and in fog, and sonar equipment to check the depth of the water.

The engine is controlled from a clean and quiet control room. Automatic instruments measure the temperature in the holds and other parts of the ship, and signal a warning if anything is wrong. In fact everything possible is done to make the tanker's voyage a safe one. But tankers are awkward ships to control. They take an enormous amount of space to turn, or to slow down and stop. Accidents can be disastrous, so smaller, more easily handled tankers are being built nowadays.

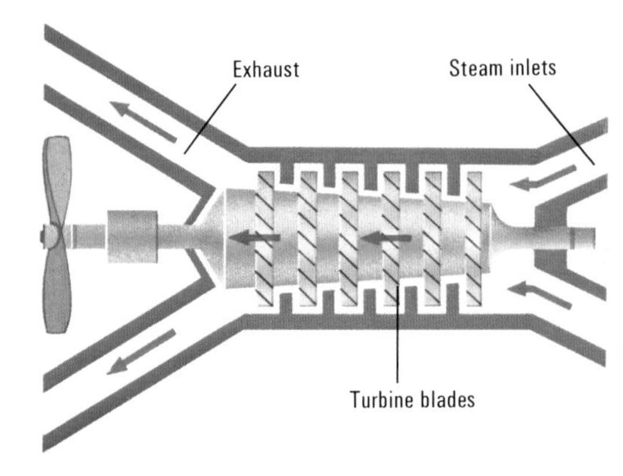

How a steam turbine engine works. A jet of high pressure steam hits and turns blades fitted around a shaft attached to the propeller.

The chart room. A chart is a sea map. It shows landmarks and other details which help the navigating officer to find the way safely. It also shows the depth of the water at high and low tide, and hidden underwater reefs and sand banks. Such information is especially important to tankers, which need very deep water.

Left: The engine control room. Complicated electronic instruments check that the engine is running smoothly and warn the officer if anything is wrong.

Above: A tanker's engine room.

ENGINE AND "BRAKES"

Most cargo ships have diesel engines, but tankers have steam turbines—which are cheaper and less complicated. Water is heated in enormous boilers to make steam. The steam comes out in a high-speed jet which hits and turns the turbine blades—just as wind turns a windmill, or water a water wheel. The turbine spins very rapidly. But the propeller which pushes the ship along only works well when turning quite slowly. So reduction gears are fitted between the turbine and the propeller.

Ships have no brakes. To slow down the propeller is reversed, so that it tries to move the ship backward. The captain must think well ahead, as it takes about 20 minutes to stop—and in that time his ship may travel up to 5 miles.

TANK CLEANING

Tank cleaning is one of the main jobs at sea. It is done when the ship has discharged its cargo and is returning to the terminal for another load. Great care is taken to let no oil spill into the sea. The top picture (A) shows two tanks (2, 3) being cleaned by swirling jets of water. The oil is washed to the bottom and pumped along a pipe to the slop tank (1). The cleaned tanks are filled with sea water (B 2, 3). In the top picture two tanks (A 4, 5) are full of water with slops on top. The water is pumped into the sea (B5), and the oil goes to the slop tank (B1). Finally (C) all the oil slops are in the slop tank on top of a layer of seawater (C1). When the tanker reloads, the clean seawater under the slops is pumped out (D), and fresh oil loaded on top of the slops. Another system uses crude oil instead of seawater for washing.

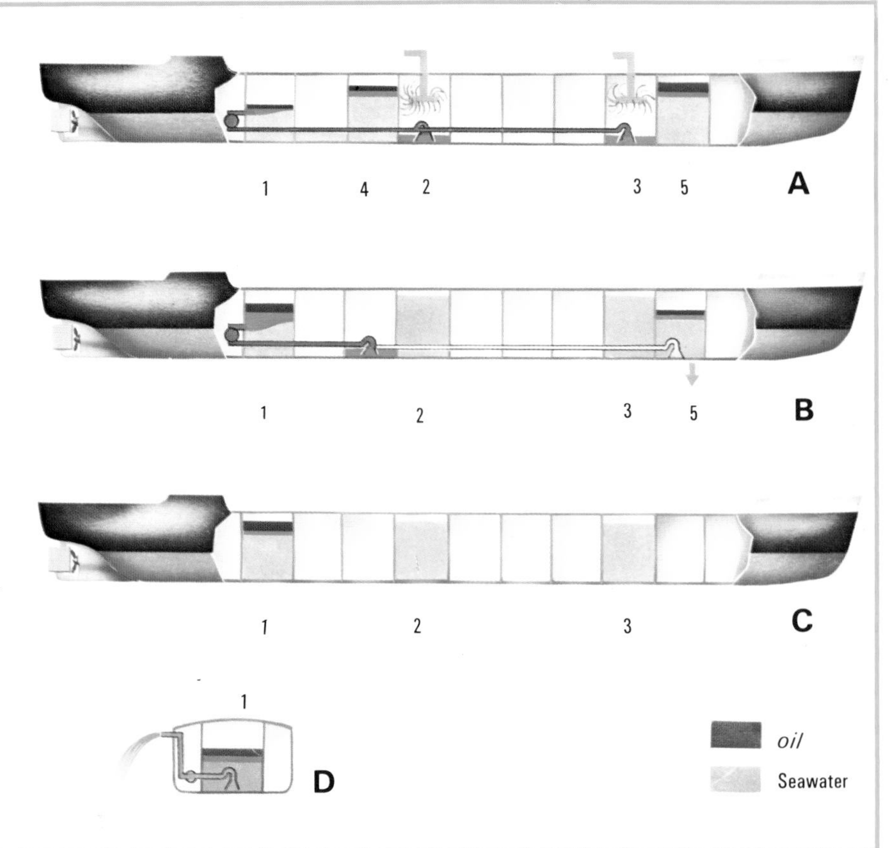

Life Aboard the Tanker

Tankers transport crude oil across the oceans, and the voyage from the oil terminal to a refinery on the other side of the world may take a long time. Today almost everything is planned and worked out by computers and other electronic instruments, and many jobs are done automatically. The chief engineer no longer has to go around feeling the temperature of pipes, listening to the "tune" of his engine, oiling this, greasing that, and tightening up the other. Instead he spends most of his time at a control board, pressing switches and watching dials. He must of course know all about the engine and its control systems, so that if anything does go wrong he can put it right. His job is vitally important, and so are those of the rest of the crew. Their ship and cargo are immensely valuable, and—if not well managed—immensely dangerous. Sometimes tankers may pass through waters made dangerous by wars or terrorism.

The ship is clean and comfortable. On many tankers every member of the crew has his own cabin, and he may spend some of his spare time there studying. Running a tanker needs much special scientific knowledge—about electronics, automatic control systems, engineering, radio and navigation, and about crude oil and its products and how to look after them safely.

The quarters are air-conditioned, the food is good, there are regular film shows, and there is a library, a games room, a swimming pool, and even a hospital. Officers are allowed to take their wives to sea every so often.

Opposite, top: 1. A member of the crew in his cabin.
2. Officers on the navigating bridge. While sailing across open sea the ship is held on course by an automatic pilot. But in coastal waters or busy seaways, or when approaching a port, the navigating officer has a difficult task controlling the speed and course of his unwieldy ship. The largest tankers are too big and too deep in the water to berth at most docks and harbors. They load and unload at an offshore terminal or buoy.
3. The officers' dining room or "mess" is roomy and comfortable, and the food is good.
4. The crew enjoy time off in the pool.

Opposite, bottom: 5. The engineer pays a visit to the engine room. He wears ear muffs, because it is so noisy there.
6. During a long voyage the tanker does not call in at ports along the route. But when it sails close enough to land, helicopters fly out with letters, papers and any urgent supplies.
7. A tanker, like a factory, has a busy office where all the paperwork is done.
8. Painting is one of those jobs that is never finished.

1
2
3
BRITISH CONQUEST
POST.
6
7

Above: The tanker's cargo pump control room with its complex array of levers and dials.

The Oil Refinery

Crude oil is greenish-brown or black, and thick and sticky. It is a mixture of many things and is not much use until all its parts have been separated at the refinery. When you boil water, it turns to steam. When the steam cools, it turns back to water—but it is then pure *distilled* water. Different liquids boil at different temperatures. At the refinery, crude oil is boiled at the bottom of a *fractionating column*. The vapors from all the different substances (or "fractions") in the oil rise up the column and cool down at different heights.

Left: When the tanker arrives at the refinery, large pipes are fixed to outlets on deck and the job of pumping the oil ashore begins. The whole job is often planned and controlled by computer. This ensures that different kinds of oil do not get mixed. It also safeguards against the ship becoming unbalanced for, if all the oil was pumped from the front tanks first, the bow of the ship would rise. Sometimes the entire load is discharged at one refinery; sometimes the tanker only unloads part of the cargo and then sails on to discharge the rest elsewhere.

As the vapor from each fraction cools to the right temperature, it turns back to liquid. Fractions which liquefy (turn back to liquid) at the coolest temperature do so high up the column, where the vapor has had longer to cool. Those which liquefy at higher temperatures do so near the bottom. So each fraction turns back to liquid at a different level, and can be collected and piped into storage tanks. The crude oil has been distilled into a whole range of products, from gas at the top to bitumen at the bottom.

Another process, called *cracking*, is used to turn some of the heavier fractions into high-grade gasoline.

RUNNING OUT OF OIL?

In 1950 the world used about 11 million barrels of oil a day. Today we burn up about five times as much. Nobody knows exactly how much oil is left. Some experts think that if we go on using so much there may be none left in 50 or 100 years. Many people think it is a waste to use oil for heating and electricity generating. We should use coal, gas, nuclear power, or energy from the sun, the wind or the waves. They say we should keep the oil for cars, aircraft, and other forms of transport. Other people believe there is enough oil under the ground to last for a long time. Still others say we should learn to make new fuels from plants.

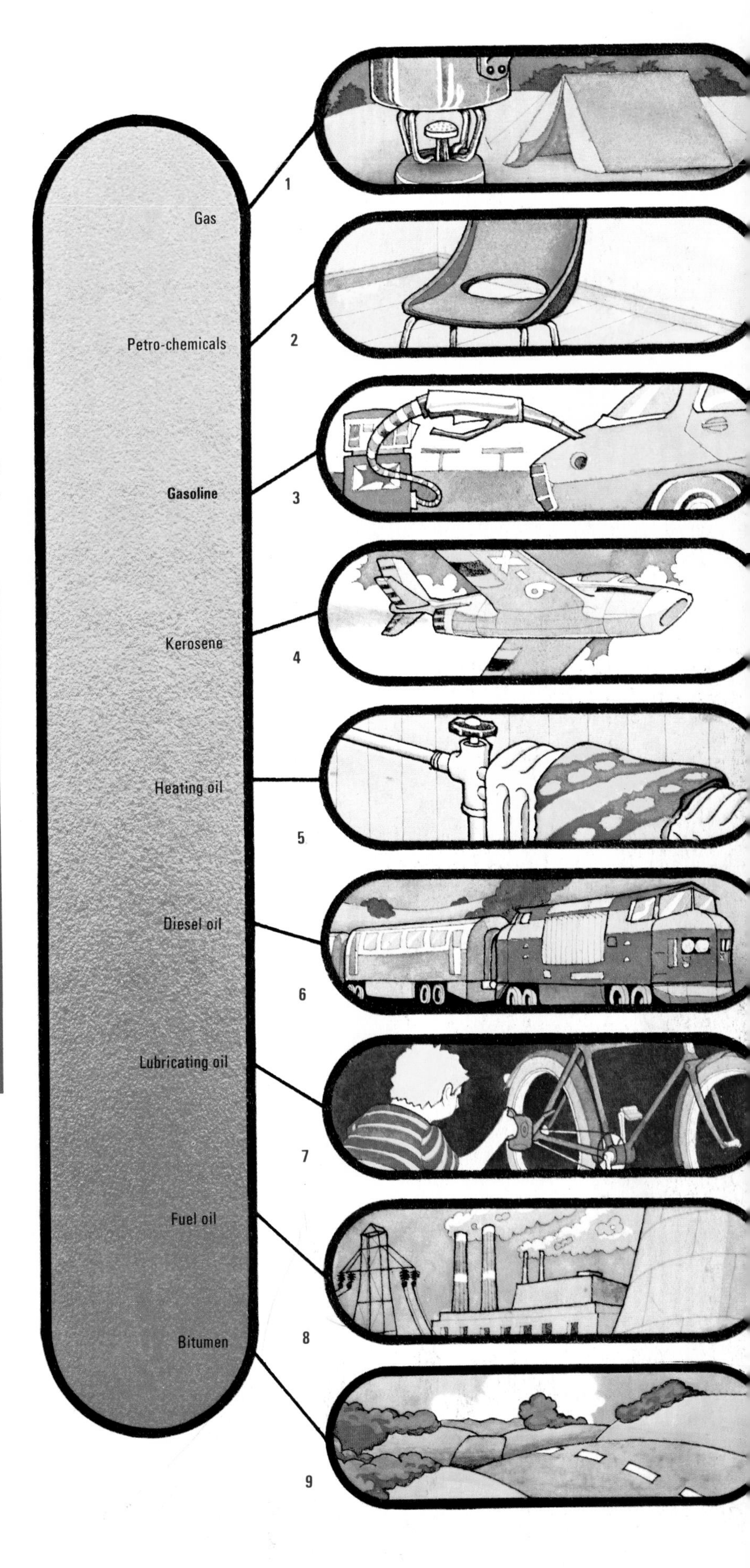

Right: The diagram shows some of the oil products made at the refinery, and some of their uses. The "lightest" fractions are at the top, the heaviest at the bottom. 1 Bottled gas for camping cookers and lamps. 2 Oil products called petro-chemicals are used for making many plastics (for example, the plastic chair in the picture, nylon for clothes, and much else). 3 Gasoline. 4 Kerosene, or paraffin, for jet plane engines, and for heating homes. 5 Heating oil for central heating boilers. 6 Diesel oil, for trains, buses, trucks, and diesel ships. 7 Lubricating oil. 8 Fuel oil, a fairly cheap oil used for steam turbine generators in electricity power stations and steam turbine engines in large ships such as tankers. 9 Bitumen, a thick, tarry substance used in road-making and in roofing felt and other waterproofing.

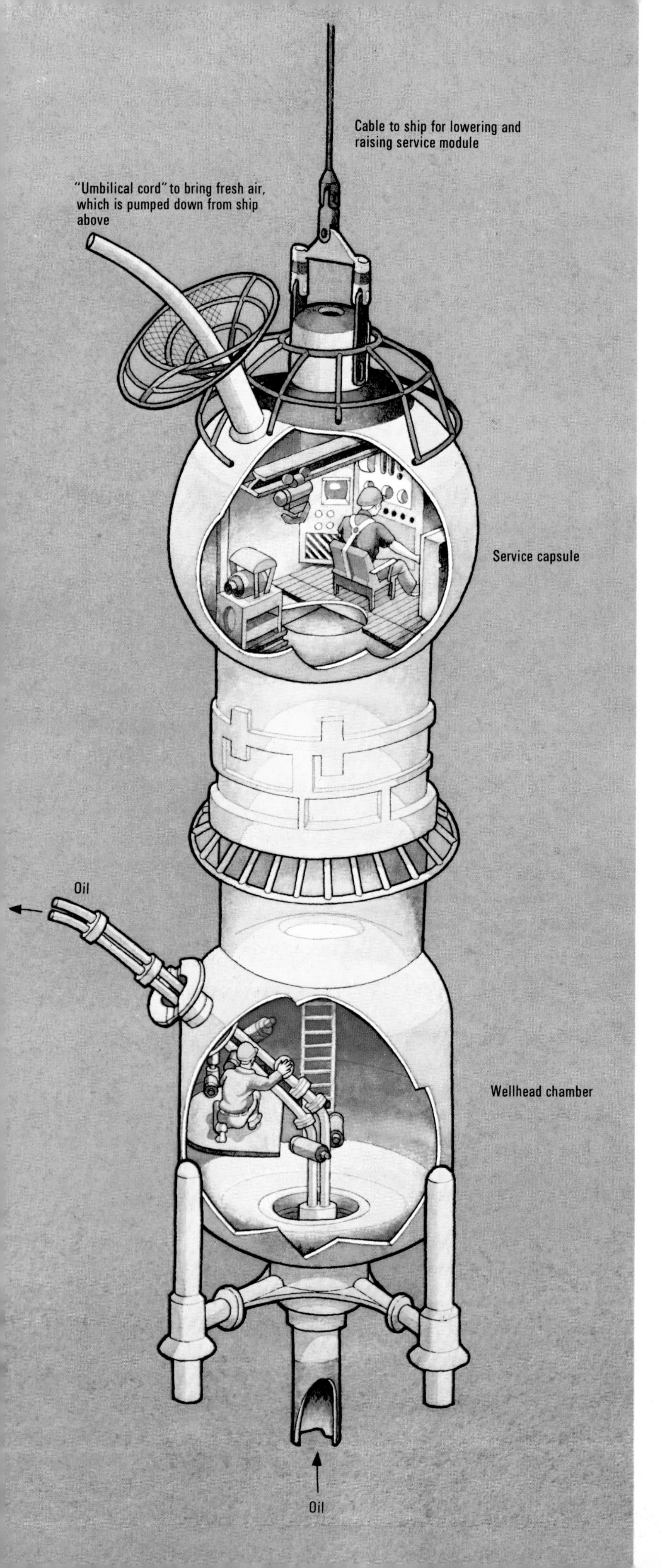

Oil Under the Oceans

Tankers often pick up oil from a buoy some distance from shore. In the future they may have to go out into the oceans, as oilmen learn to drill in even deeper water. The deeper the water, the more difficult it is to drill, and the more expensive it is to set up a production platform.

Already some wells are producing oil without a production platform. The picture on the left shows how this could be done. One day all the work, except for the drilling, may be done from chambers on the sea bed. You can see what it might look like in the large picture. At the top left is a floating drilling rig. The drill string passes down through a fixed chamber similar to the one at the bottom of the picture. In this and all the other seabed chambers the men live and work just as they would on a surface rig. When the well is complete, the *Christmas tree* that controls the flow is fitted inside the chamber (right hand end). The equipment used to treat the oil is there too (in the middle), ans so are the living quarters. Small submarines bring supplies, mail, and workmen.

The processed oil is piped from each production chamber to the module in the middle of the picture. This is made up of several *cells*. Most are for oil storage, but some are for further processing, and one provides more living quarters. Waste gas is piped from here to the flare at the top, while the oil is piped up to the floating buoy and storage unit (top right). From here it is taken ashore by tankers.

Left: If a field is too small to make it worth building a platform, or in too deep water, oilmen set up a fixed "wellhead chamber" on the seabed. Inside it men can work just as they would on a surface rig. They need no diving gear or breathing equipment. To reach the chamber they are lowered from a ship in the "service capsule." This locks onto the wellhead chamber.

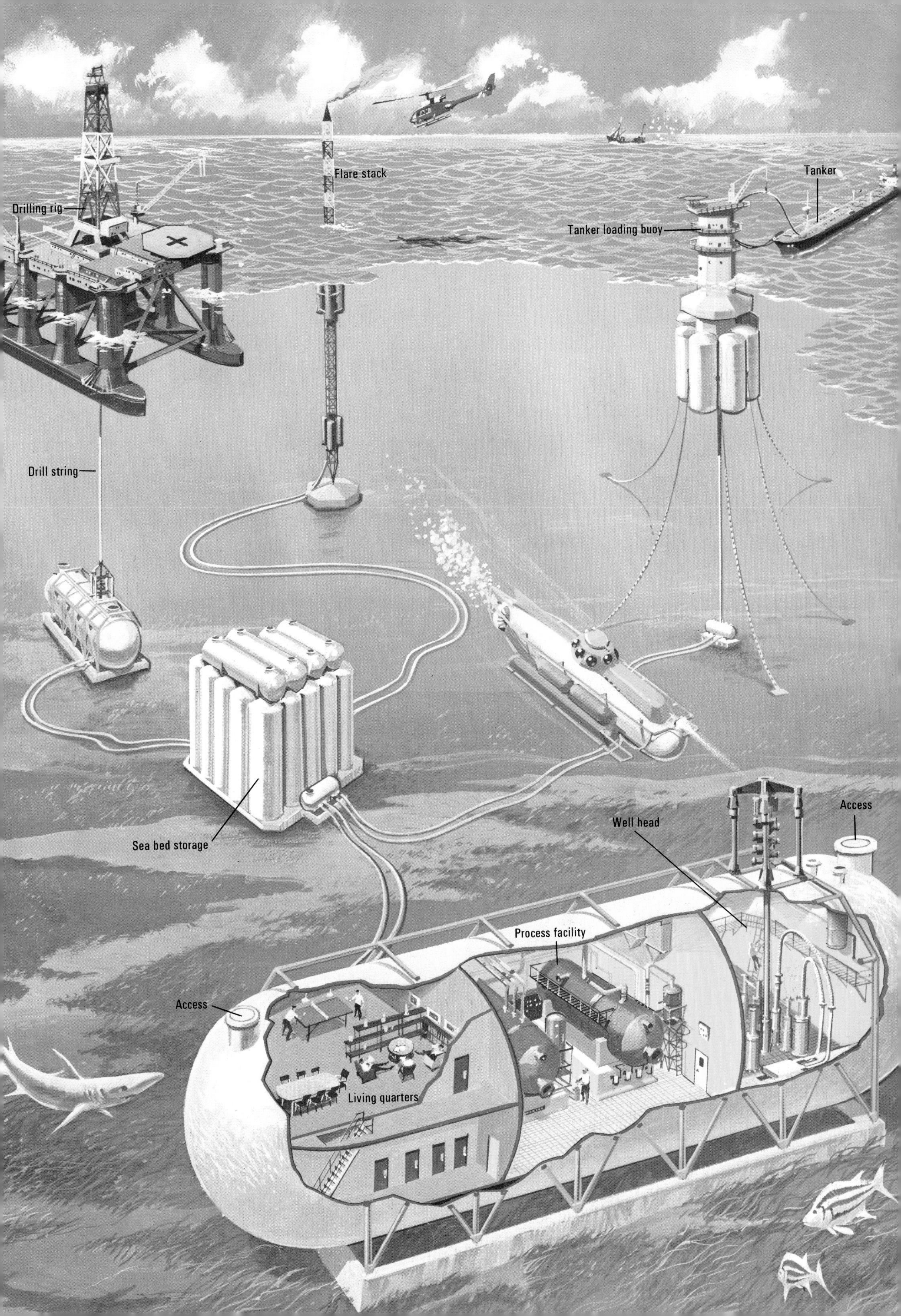
Flare stack
Tanker
Drilling rig
Tanker loading buoy
Drill string
Access
Well head
Sea bed storage
Process facility
Access
Living quarters

Tankers Around the World

Oil deposits have been found in almost every corner of the world, and the search is still going on. Tankers and the other vessels used in the transport and discovery of oil travel thousands of miles from place to place in conditions that can sometimes be dangerous.

As well as the weather and sea conditions, the crew must always be alert to the danger of a fire breaking out. More recently, new dangers have appeared. The threat of terrorism and war, particularly in the Middle East has created problems which have had an impact all over the world.

Because of their sheer size, tankers are very slow to maneuver, and as a result they must depend on other vessels, either to protect them against attack, or to help them make their way through busy shipping lanes when they are entering or leaving ports.

Above: Although it is shown here trapped in the Arctic ice, the oil tanker Manhattan *managed to make the Northwest Passage in 1969.*

Left: Fire is the most dangerous hazard to oil rigs and tankers. Special emergency ships carrying powerful fire-fighting equipments are available to tackle a disaster at sea.

Above: Seen from the air, a supertanker dwarfs the tugs that help it into its berth. The sheer size of oil tankers makes them difficult to maneuver, particularly in busy inshore waters.

Right: The Persian Gulf is a vital international waterway, used constantly by oil tankers. During the Iran-Iraq War, this route has become dangerous and warships, such as this British frigate, patrol the Gulf so as to escort tankers.

IMPORTANT HAPPENINGS

1859

1859 The world's first successful oil well is drilled, in Pennsylvania.
1885 The first successful gasoline engines are built by the German car engineers Daimler and Benz.
1886 *Gluckauf,* the first ocean-going oil tanker, is launched. It carried around 3,000 tons of oil. The first giant "supertankers" were built in the 1960s.
1891 Oilmen prospect for oil off the coast of Southern California. They drilled from piers sticking out from the shore in shallow water.
1920s First offshore oil drilling in coastal swamps of Louisiana.
1935 The first "drilling barge" is built. It was towed out and sunk in position, resting on the seabed.
1949 The first mobile offshore drilling rig is built. It could only be used in shallow water.

1950s

1954 The first mobile jack-up rig begins drilling in the Gulf of Mexico. At about the same time the first floating rigs were tried. They paved the way for the huge "semi-submersible" (partly sunk) exploratory rigs of today.
1959 Gas is found under the sea in Slochteren, the Netherlands.
1964 First gas field found off the east coast of England.
1969 Oil discovered in Alaska
1969 Oil tanker *Manhattan* succeeds in making the Northwest Passage through the Arctic ice.

1970s

1970 In June, Norway discovers the first giant oil field in the North Sea, the Ekofisk field. In October Britain makes her first major find, the Forties field.
1973 Countries belonging to the Organization

Below: The disused and rusting tankers shown below in a Norwegian Fjord are a reminder that the total world demand for oil has dropped in recent years.

of Oil Exporting Countries (OPEC) reduce oil output, with the result that the price of oil increases fourfold. This leads to an "energy crisis" in the industrialized world.

1975 In June Britain's first North Sea oil comes ashore, from the Argyll field.

1977 Trans-Alaskan oil pipeline begins operation, carrying oil from Prudhoe Bay to the ice-free port of Valdez.

1979 Launch of the world's largest oil tanker to date, the *Seawise Giant*.

1980s

1980s Movement of oil through the Persian Gulf is disrupted by the war between Iran and Iraq. Several oil tankers are hit by missiles or mines.

1981 Work begins on a 2,760-mile pipeline to carry natural gas from Siberia, U.S.S.R., to Western Europe.

1985 World demand for oil falls to a level below that of 1973. Unwanted tankers are scrapped, and world tanker fleet at total of 224 million tons falls well below 1977 peak of 301 million tons.

Right: A tanker loading up at Valdez, Alaska.

DISASTERS

1967 The *Torrey Canyon* ran aground off the southwest coast of England. Oil polluted a long stretch of coast.

1976 The Norwegian tanker *Berge Istra* was ripped apart by explosions and sank within one minute.

1978 The *Amoco Cadiz* ran aground near Portsall in France. More than 200,000 tons of oil were spilt into the sea and onto coastline.

1983 The tanker *Castillo de Bellver* caught fire off Cape Town, South Africa: over 225,000 tons of oil were lost.

1983 What may have been the world's worst oil spill happened at the Nowruz oil field in the Persian Gulf, after a blowout. More than 500,000 tons of oil spilled into the sea.

GLOSSARY OF TERMS

Ballast man Rig crewman who makes sure the ballast in the pontoon of a floating exploration rig is correct to keep the rig level in the water.

Barrel Unit used to measure oil. One barrel equals 160 liters (42 U.S. gallons).

Bends Sickness that affects divers if they "decompress" from working in deep water too quickly. Caused by bubbles of nitrogen forming in the blood.

Bit Cutting tool used to drill through mud and rock.

Blow-out Escape of oil or gas during drilling. A hydraulic blow-out preventer at the well head stops such accidental escapes of oil.

Casing Steel pipe cemented into a well to line the hole.

Christmas tree Name given to the complicated assortment of valves and pipes at the well head. They control the flow of oil.

Coring Taking a rock sample from a well.

Crude oil Oil in its natural state, usually underground.

Decompression After working underwater, where water pressure is great, a diver must be returned slowly so that his body adjusts back to normal atmospheric pressure. This is done inside a special decompression chamber of the diving boat.

Derrick Tower which stores the lengths of drill pipe.

Diving bell Chamber lowered into the sea from which divers can inspect underwater equipment and carry out repairs.

Drill-mud Liquid used to cool and lubricate the drill-bit, which would otherwise overheat or clog. The drill-mud helps to line the sides of the well, and to bring rock and rubble back to the surface.

Dry dock Large dock that can be drained of water. Used to build tankers and other large vessels, and to carry out repairs.

Duster An empty well, or a "dry hole."

Flare stack Tower through which unwanted gas is allowed to escape and is burned off.

Fossil fuels These include coal and oil, so named because they were formed millions of years ago by the fossilization of tiny plants and animals.

Gravity platform Rig that is held in place by its own weight.

Gusher A sudden escape of oil or gas, gushing high in the air. Such escapes often happened in the early days of oil drilling, but are avoided today by using blowout preventers.

Hydrocarbons Chemical compounds of hydrogen and carbon. Oil and natural gas are composed of mixtures of naturally occurring hydrocarbons.

Right: An oil rig silhouetted against the sky, showing two flare stacks through which unwanted gas can be burned off.

Jacket The steel tube structure that stands on the sea bed and supports a production platform.

Jack-up rig Rig used in shallow waters. Its legs are jacked down to rest on the bottom while drilling is going on.

Kick A well "kicks" if the pressure of oil or gas is greater than that of the "mud" column.

Killing a well Controlling a wild well or "mudding off" a completed well to hold the reservoir pressure in check, by filling the borehole with "mud."

Natural gas The gas which is produced naturally with oil.

Oil sand Porous sandstone containing oil.

Pay zone Rock in which oil and gas are found in large enough quantities to be worth exploiting.

Petrochemicals Chemicals made as by-products from oil or natural gas.

Pontoon Floating structure used to support a semi-submersible rig, and which can be raised or lowered by altering its ballast.

Roughneck Name given to member of crew of a drilling rig.

Seepage A natural oil spring which occurs at the surface.

Spudding in Starting drilling a new hole.

Support vessels These include firefighting craft, and diving support ships.

Semi-submersible Rig that is kept stable in the water by floating pontoons. Used for deep-water drilling.

Tool pusher Person responsible for drilling work on a rig.

VLCC Stands for Very Large Crude Carrier, another term for a supertanker.

Terminal Point at which tankers load and unload cargoes.

Left: Roughnecks work in dirty and slippery conditions during drilling operations.

Below: Divers are very important in the day-to-day work on an oil rig.

INDEX